Sex Differences
- A LAND OF CONFUSION -

SEX DIFFERENCES
A LAND OF CONFUSION

a look at the scientific literature on
the average differences between males and females

ZACHARY ELLIOTT

© 2017 Zachary Elliott. All rights reserved.
ISBN 978-1-387-38091-6

TABLE OF CONTENTS

Preface vii

A Land of Confusion 1

    Gender Inequality across Western Liberal Nations 4

    Objectives of this Paper 9

    Gender as a Social Construct 10

    A Lesson on Bell Curves 17

    Average Sex Differences 19

        Differences in Cognition and the Brain 24

        Differences in Personality 30

        Differences in Interests and Occupational Preferences 39

    Average Sex Differences: Summarized 43

Conclusions 48

References 54

Preface

In the summer of 2017, a software engineer from Google named James Damore circulated an internal memo at the company, discussing why Google's uneven amount of men and women in their tech sector was not due to discrimination, using scientific literature from the fields of psychology and biology to explain his point. The disparity, he argued, was mainly due to average biological differences between men and women in their psychology and personality, leading them to choose different career paths.

Opening his memo by declaring, "I value diversity and inclusion," Damore presented methods he thought would be effective at increasing the number of women at Google, without resorting to quotas or discriminatory hiring practices. For a few weeks, everything seemed fine, as the memo circulated privately around the company, with a large portion of Google workers agreeing with it. All of that changed when the memo was leaked to a news source. Within just a few days, social media exploded, with news organizations dubbing it the "Anti-diversity Memo", quoting Damore as saying that "women were unsuited for tech jobs" even though he never wrote that. A few days later, Damore was fired from Google. It didn't matter that his opinion was backed up by decade's worth of mainstream scientific research--research that has been widely and publicly available for decades, with most psychologists, biologists, and evolutionary behavioral scientists supporting it. All that mattered was that he dared to disagree with what he called "Google's ideological echo chamber."

That summer, I was following the story closely as it developed. I had been interested in psychology and personality research for years, and the literature on the differences between males and females was fascinating. On scientific journals, I'd find myself reading about the average differences between the sexes. And I was curious. I loved to learn about how our brains work, how our bodies work, and how our decisions are shaped by our psychology, interests, and desires. There's something beautiful about understanding the complexity between males and females--the beauty of realizing how we can be very similar and very different simultaneously. *Very similar in many aspects, very different in others.* When the memo story unfolded, I immediately read Damore's memo to see if the headlines were true. *Did he actually say women were unsuited for tech? Did he even imply it?* I was pleased to find that the memo, an evil anti-diversity screed in the minds of the media, was actually very balanced, reasoned, and ultimately, centrist. It was a well-written, articulate paper which clearly explained the mainstream scientific literature on average sex differences, and I was impressed with how concise and straightforward it was. Furthermore, he presented ways of increasing women at Google through understanding these average differences.

For years, I had been hearing that there were no inherent differences between males and females, and that any differences were merely social constructions--completely contrary to the existing evolutionary literature on sexual dimorphism in mammals. I was being told, like most people in our society, that disparities between the sexes in occupations, for example, are evidence of injustices. The memo, and the hysteria that encompassed it, inspired me to do some research, and ultimately...write. I learned many things through my research: I learned that many sex differences we see have major biological causes; that many of the disparities we see in occupations, contrary to the prevailing views, are not the result of sexism or discrimination, but of people's choices and preferences; that this topic, the degree to which biology and society affect us, is very complex and nuanced with interconnected variables and many layers of analysis. Most importantly, however, I learned each individual is incredibly variable and unique, a testament to the importance of valuing the individual over group identity.

As a way to refine my arguments and further develop my knowledge on the subject, this paper was written as a concise but nuanced overview of the scientific literature on average sex differences. The evidence, as it turned out, was insurmountable.

A Land of Confusion

In July 2017, *HuffPost* ran a story about a female-to-male transgender artist named Cass Clemmer who posted a photo of themselves sitting on a bench with their legs spread open, showing menstrual blood soaking their pants. The sign Clemmer was holding read, "Periods are not just for women. #BleedingWhileTrans."[1]

Clemmer, who prefers others to use *they/them* pronouns, told *HuffPo*, "Not all people who menstruate are women, and not all women menstruate," yet is that true? Fundamentally, how you view this statement depends on how you *define* a man and a woman. Does the word *woman* mean an *adult human female?* If that's the definition we are using, the first part of Clemmer's statement is nonsense. If the word *woman* means an *adult human female*, then the statement, "not all people who menstruate are women" is incorrect. Menstruation, the discharge of blood and tissue from the inner lining of the uterus through the vagina, can only occur in biological females. Therefore, if the word *woman* means an *adult human female,* then it is safe to assume periods only happen to women. But clearly, this is not what Clemmer means. They are using a different definition of *woman,* removing its biological definition completely and replacing it with an

---

[1] Nichols, J. (2017). Women Aren't The Only Ones Who Get Periods. *HuffPost.*

entirely cultural one. In Clemmer's view, the word *woman* is defined as an adult who ascribes to themselves feminine characteristics, regardless of their biological sex. You may have a biological male who identifies as a woman, or a biological female who identifies as a man. With that definition, the last part of Clemmer's statement, "Not all women menstruate," now makes sense. There are many biological males, specifically transgender individuals, who identify as a woman, and there are many biological females who identify as a man. Therefore, *not all women menstruate.*

Clemmer claims that anyone who disagrees with their statement is transphobic, but transphobia is defined as a range of negative attitudes, feelings, or actions toward transgender or transsexual people. And many who objected to the artist's statement never acted in any transphobic way. There was simply a principled disagreement with Clemmer on how to define *man* and *woman.*

But why the misunderstanding? Why is there a disagreement on what should be a clear definition? It fundamentally comes down to a person's views on the interaction between sex and gender. For decades, we've known that sex and gender are not entirely the same. Sex is based on a person's reproductive functions, their genetic makeup. Gender is based more so on social and cultural differences and an individual's identity. The degree to which sex and gender interact continues to be heavily debated. One view is that sex and gender are completely separate and malleable entities, completely molded and shaped by society and culture, and these entities do not interact with each other. In this view, a person's sex has no effect on their identity, meaning there is no correlation between sex and gender, and therefore, gender is only shaped by society. A second view is that sex and gender are so tightly linked that there is no difference between the two. Your sex is either male or female. Your gender is either man or woman. In this view, a person's sex directly affects their identity, and therefore, gender is only shaped by biology. However, both views tend to oversimplify what is a much more complicated and nuanced debate, which is that sex and gender affect each other in differing degrees across different traits. While they are not completely separate, they are not completely the same either. In fact, the interaction between sex and gender stems from a multitude of complex and interconnected variables, with biological, societal, and environmental factors.

The popularity of Cass Clemmer's argument in today's society provides insight into the different views of how sex and gender are seen, leading us into a discussion about the very nature of sex and gender itself. Clemmer's argument that *periods are not just for women* is ultimately a social constructionist claim based on a different definition of the word *woman*, disregarding biology altogether and blurring the lines between male and female. It symbolizes a larger discussion happening in our society regarding the differences between men and women and whether these differences are the result of social constructions or evolutionary dimorphism.

There is a lot of debate among today's youth about what sex and gender are, how they are connected, and the differences between the sexes, and there is a lot of confusion being spread: that there are no differences between the sexes; *no differences in their psychology, no differences in their interests or desires,* and *no differences in their behaviors;* that any apparent differences we think exist are simply the result of social constructs, a natural product of a patriarchal culture, and that to claim otherwise is sexist and misogynistic; that gender is only a social construct with no basis in biology; that gender differences only arise through socialization, and that disparities in occupations are evidence of sexism and discrimination; that a woman at home taking care of her children is a product of sexism, while a woman focused solely on her career is enlightened and empowered; that women aren't the only ones who can have periods and become pregnant, and that men can do it too.[2] Like the name of the 1986 song by *Genesis*, we are living in a *land of confusion*.

Gender Inequality across Western Liberal Nations

Sexism and misogyny, unfortunately, were parts of our society in the past. Women were held back from the right to vote, the right to own property, and the right to work in the same job as a man. Sexism certainly still exists in the form of individuals, and where claims of sexism or discrimination based on sex or gender arise, clear evidence of discrimination must be provided. Forms of bias also still exist among individuals, favoring men over women in certain situations and favoring women over men in others. Yet today, most Western nations have liberalized and adopted mostly free-market systems that allow anyone, of any race, sex, or gender, to pursue their interests. While these Western liberal nations have become very egalitarian, pushing policies that support gender equity in occupation, pay, and hiring, we still see large inequalities in many things. Consider a few of these inequalities:

---

[2] Coleman, N. (2017). Trans Man Gives Birth to Baby Boy. *CNN*.

1) The difference in percent of U.S. doctoral degrees awarded to men and women in specific fields is larger in some and smaller in others, as seen in figure 1. Doctoral degrees are heavily dominated by women in fields such as health sciences, education, public administration, social/behavioral sciences. The percent of doctoral degrees in the health sciences awarded to men is only 30% (70% of these degrees are given to women). The numbers for freshman enrollment in these fields are almost the same.

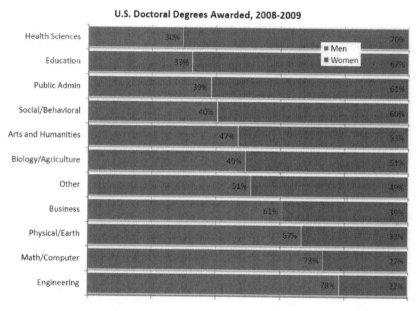

Figure 1. *U.S. Doctoral Degrees Awarded by field and sex.*

2) As of 2013, 60% of U.S. bachelor's degree holders are women.[3] And 57% of enrollees in graduate programs are women.[4]

---

[3] Fisher, A. (2013). Boys vs. Girls: What's Behind the College Grad Gender Gap? *Forbes*.
[4] Wang et al (2013).

3) The percent of bachelor's degrees earned by women in psychology, the social sciences, biosciences, and the physical sciences has been steadily on the rise since 1991. The bachelor's degrees in Psychology, social sciences, and biosciences all have a female majority. Percent of bachelor's degrees earned by women in computer sciences and mathematics have declined slightly since 1991, with men being a majority in the computer sciences and engineering, and only a slight majority in mathematics.[5] (See figure 2).

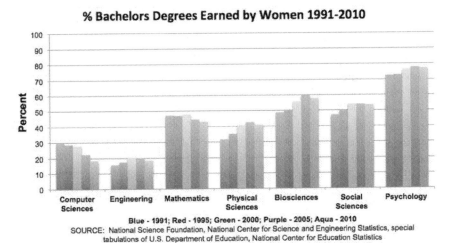

Figure 2. *Percent of Bachelor's Degrees Earned by Women, 1991-2010.* National Science Foundation, National Center for Education Statistics.

4) The United States Department of Labor collects data on the percent of the population working in certain jobs. Dental hygienists are almost exclusively women (98% women), while brick and cement masons are *exclusively* male (0% women). Women make up large majorities as therapists, licensed and vocational nurses, registered nurses, elementary school teachers, and mental health counselors. Men make up large majorities as carpenters, mechanics, truck drivers,

---

[5] Source: *National Science Foundation* and *National Center for Education Statistics.*

software developers, and sales representatives. Percent of men and women is more equal in physician assistants, accountants, marketing specialists, medical scientists, postsecondary teachers, database administrators, and physicians.[6]

5) Taking all the *average* earnings of men and women, women make about 80 cents to every dollar a man makes. The wage gap is one of the most cited statistics supposedly proving that companies are discriminating against women (i.e. paying them less for the same work).

Yet this statistic simply takes the *average of all earnings*. It doesn't account for the type of job or hours worked and many other factors that contribute to the difference in the two averages. When these factors are taken into account, the wage gap is reduced to only a few percentage points of the 20 percent gap.[7] Of course the statistic is not equal, because women tend to go into jobs that pay less (such as the social sciences) or they work less and decide to take time off to care for children. It is true that a very small percentage is likely due to forms of discrimination, but it is illegal in Western nations to discriminate someone on the basis of sex, race, gender, religion, sexual orientation, or any other form of identity. For example, if an equally experienced woman and equally experienced man are working the exact same job with the exact same hours, it is illegal to pay them differently based on sex.

Even though economic opportunity, economic mobility, and freedom of choice are very high in these liberal nations, disparities such as these continue to exist between men and women. And gender sociologists continue to claim this is because of discrimination or the

---

[6] Source: *United States Department of Labor.*
[7] Horwitz, S. (2017). Truth and Myth on the Gender Pay Gap. *Foundation for Economic Education.*

patriarchy, without supplying much evidence other than showing that the disparity exists. For decades now, quotas in hiring have been instituted to try and bring equal representation of women in certain fields, especially STEM, with varying degrees of success. For example, the percent of Computer Science bachelor's degrees held by women fell from 30% in 1991 to 19% in 2010.[8] Has sexism in that field increased since then? Or could these disparities simply be a product of men and women's choices and inclinations?

What if equal representation in *every* field (50% men and 50% women) is unreasonable or even impossible if liberal democracies give people the freedom to make their own choices?

While we can agree that gender *equality* (equal opportunity for the sexes) is a moral and just cause worth fighting for, there are those, especially gender sociologists, who believe we must have gender *equity* (equal outcomes and no disparities between the sexes). Is this utopia of perfect gender equity something we should be pushing for? The social sciences, for example, have a large gender gap--a gap favoring women--while STEM fields tend to be heavily populated by men. Does that mean the social sciences are sexist against men, or is it that men, on average, are less interested in those fields? Most prison inmates are men--93% to be exact.[9] Does that mean the criminal justice system is steeped in sexism against men? Or do men commit most of the crime?

What if these differences could be explained through the freedom of men and women to make their own choices in liberal democracies?

---

[8] Source: *National Science Foundation* and *National Center for Education Statistics*.
[9] Carson, A. (2013). Prisoners in 2013. *Department of Justice, Bureau of Justice Statistics*.

What if, instead of resulting from the *patriarchy, sexism,* or *societal-imposed gender roles*, these differences are mostly evidence of men and women's preferences rooted in biological factors?

Objectives of this Paper

1) To address the claims of social constructionists by showing that **average sex and gender differences between human males and human females are real**, and that many of these differences have a strong basis in biology.

2) To show that the **inequalities seen across society in things such as occupation can be mostly explained through a multitude of biological and psychological factors** (i.e. that women and men have, on average, differing interests).

3) To **reinforce the idea that there is significant overlap between males and females in many traits**, and that individuals should be judged as *individuals*, not members of their group. And that our similarities are more important than our differences.

The changing definitions of what constitutes a man and what constitutes a woman is emblematic of a deeper discussion between the fields of sociology, evolutionary biology, and psychology regarding the roles of biology versus environment--the age-old argument regarding the balance between nature and nurture, and the degree to which these forces affect our behaviors relating to sex and gender. So what's the scientific consensus? Are the sex and gender differences between men and women due to biology or are they due to society? What about a mix of both? The degree to which these factors affect our behavior and the conclusions we draw from it have large ramifications for our society's medical, mental, and sexual health. Before we answer these questions, we need to examine the views held by the majority of today's gender sociologists.

Gender as a Social Construct

Most gender sociologists claim that there are no inherent differences between males and females in their psychology, behaviors, desires, and interests. Most or all differences between males and females observed in society are social or cultural constructions with almost no bearing on biological factors. In this sense, the term *gender* is defined as the social constructions of masculine and feminine traits, a view perfectly summarized by feminist and social constructionist Simone de Beauvoir in the 1940s, "One is not born a woman, one becomes one." In de Beauvoir's view, everyone is born *tabula rasa*--a blank slate--and any apparent differences arise through the mold of society. Rejecting biology altogether, many social constructionists even claim that we are, in fact, not born as anything.[10] All the choices we make, all the things we do, they are all affected by the environment or the societies and cultures in which we live.

Therefore, in this view, it makes perfect sense for Cass Clemmer, our transgender artist, to claim that women are not the only ones who can have periods. If gender is simply a social construct and we are all born with no biological predispositions in regards to our sex, then those who claim to be men can have periods too.

There is one major philosophy that has caused many of these culture shifts for the past few decades, and that philosophy is

---

[10] Holmes, M. (2007). *What is Gender? A Sociological Approach*. Sage Publications.

Postmodernism. Beginning as a response to the apparent failures of Marxist socialism in the middle of the 20th century, Postmodernism offers a unique critique of the current capitalist system across liberal nations, similar to the critique made by Karl Marx in the 19th century, but also slightly different. Before we move on, we must understand the principles of Postmodernism.

Much of the ideas being pushed by the gender sociologists today are heavily influenced by the work of Postmodernist philosophers like Michael Foucault, Judith Butler, Denise Riley, and Jacques Derrida, and these philosophers are all monolithically *far* Left on the political spectrum, and the reason for that is unclear. One would expect a large philosophical movement like Postmodernism to have a wide variety of viewpoints across the political divide, a diversity of viewpoints we see in philosophies like modernism, or political philosophies of liberalism, conservatism, or libertarianism.[11] Yet that is not the case. For some reason, Postmodernism as a philosophical school of thought is an exclusive phenomenon of the far Left, and this can easily be deduced by studying the history of the leading Postmodernist philosophers and their written works. Writing in the context of the collapse of the Soviet Union in 1991, Derrida said in his book, *Specters of Marx*, that it is the duty of the interpreter to preserve the spirit of Marxism into the 21st century, arguing that ideas of Marx must continue on into the new century.[12]

Postmodernist philosophers argue that the Enlightenment beliefs in reason, science, and egalitarianism have led to the creation of sexist, racist, and imperialist societies, which are still sexist, racist, and imperialist today, pointing specifically to Western liberal

---

[11] Derrida, J. (1994). *Specters of Marx*. Routledge.
[12] *Ibid.*

nations.[13] The Postmodernist philosophy views any system of hierarchy as a tyranny, a power struggle between the oppressor and the oppressed, even though most of the hierarchies between people in liberal societies today are based in competence rather than power, a meritocracy so to speak. For instance, a person may excel past another person due to more determination, skill, or luck, climbing the metaphorical hierarchy of competence--yet Postmodernists view it as unfair and unjust. *That person only achieved that position because of power*, so they say.

To try and eradicate the power struggle between the oppressor and the oppressed, the Postmodernists use a weapon, and that weapon is language. In their view, everything is constructed through language, which means words can actually shape reality. Since Postmodernism focuses on how constructs are created through words, gender differences, they argue, are produced solely through language. The very idea of a woman, say the Postmodernists, is a complete social construct with no real basis in reality.[14] This is the school of thought that says all truth is relative, but at the same time claims that liberal nations are still patriarchal and racist, that women in these nations are still oppressed, that there are no real differences between the sexes, and that any difference is merely imposed by society.[15] It follows logically that because all gender differences are socially constructed, any disparity between the sexes is considered evidence of sexism. Unfortunately, a large majority of gender

---

[13] Hicks, S. (2004). *Explaining Postmodernism: Skepticism and Socialism from Rousseau to Foucault*. Scholargy Publishing, Inc, 197; 211.
[14] Holmes, M. (2007). *What is Gender? A Sociological Approach*. 71.
[15] Hicks, 185.

sociologists follow these Postmodernist principles, simplifying an incredibly complex world into a rigid ideological echo chamber.[16]

Going further still, many radical sociologists believe that the only way to achieve liberation is to completely restructure society through the removal of the patriarchy (the ruling class), the Postmodernist version of a Marxist revolution.[17] Similar to the Marxist ideas of class struggle between the bourgeois (the owners of production) and proletariat (the workers), Postmodernism continues this tradition but alters the definitions from *class* to *identity*. Instead of the bourgeois, you have the white heterosexual male, the historical oppressor, and instead of the proletariat, you have intersecting minority groups, the victims of an oppressive society. Instead of the argument for *need* in Marxism, you have the argument for *equity* in Postmodernism. "No longer was the primary criticism of capitalism to be that it failed to satisfy people's needs. The primary criticism was to be that its people did not get an equal share."[18]

Whether you agree or disagree with these philosophies, it is undeniable that the Postmodernist ideas of relativism, social constructs, equity, intersectionality, critical theory, and deconstruction are the leading philosophical principles in the social sciences today, including the sociology of gender.

By 1980, most gender sociologists agreed that the term *gender* was to be conceived of as a sociocultural concept. Social constructionism, and the Postmodernist school of thought that encompassed it, had become the dominant theory, and as more and more social sciences became influenced by philosophers like Foucault and Derrida, a transformation began to take place in the sociology of

---

[16] Hicks, 85.
[17] Hicks, 151; 188.
[18] Hicks, 158.

gender.[19] Mary Holmes, author of *What Is Gender? A Sociological Approach*, clarifies what social constructionism implies about biology. "To say that gender is socially constructed was to resist ideas about women which assumed that differences between women and men were biological and therefore unchangeable."[20] In other words, social constructionism, in regards to gender, posited that societal and cultural constructs were the exclusive forces that shaped the differences between men and women. The logic follows that if all societal and cultural influences on sex and gender could be removed, then both men and women could live lives of perfect gender equity. Gender, as described by the social constructionists, would not exist.

The idea that men and women are shaped mostly by society has implications in both the public and private sector. For example, if there is an unequal distribution of women versus men in an organization, social constructionists, working under the assumption that gender differences are simply constructs, view this disparity as 1) *a product of sexism* and 2) *a product of societal-imposed gender roles*. The conclusion is obvious: Fix the injustice through quotas and attempt to achieve an equal amount of men and women in the organization.

As you will see, this initial assumption that a disparity between the sexes is necessarily evidence of an injustice can be easily explained away through many factors, and these factors are heavily tied to an individual's physical characteristics, cognition, brain, personality, and interests--each one affected by both biology and the environment to differing degrees, which affects the choices individuals make.

---

[19] Stephen Hicks's *Explaining Postmodernism* describes why concepts such as social constructionism are part of the larger Postmodernism philosophy encompassing the New Left.
[20] Holmes (2007). *What is Gender? A Sociological Approach*. 90.

Most sociologists and social scientists in the humanities today, especially those who study the sociology of gender, tend to fall on the nurture end of the nature-versus-nurture debate. They can be considered either social determinists or social constructionists under the Postmodernism philosophy. They believe that societal and cultural factors alone affect an individual's behavior, which is in contrast to biological determinism, the idea that genetic factors alone affect an individual's behavior. For the social determinists, the choices people make and the way they act can only be described through the society, and any disparity between males and females is seen as evidence of prejudice. Any discussion about the very relevant biological and psychological factors that contribute to *average* differences in interests and choices among the sexes is considered taboo and misogynistic.[21] An example of the logic of social determinism can be found in common claims from the social sciences, such as *poverty causes crime* and *climate change causes terrorism*. If humans are indeed simply products of the environment they are living in, then it follows that we have no free-agency. *After all*, say the social determinists, *it's not someone's fault they committed a crime. External factors caused them to do it.*

Whether it be biology or the environment, both social determinists and biological determinists agree that external factors outside of our control determine our choices. Most scholars today consider these views outdated.[22]

After decades of extensive research on the subject, most scholars and scientists arrived at a different conclusion, agreeing with neither

---

[21] Pinker, S. (2004). Why Nature and Nurture Won't Go Away. *Daedalus 133(4)*, 9.

[22] Pinker, S. (2003). *The Blank Slate: The Modern Denial of Human Nature*, Penguin Books.

the exclusive social determinism of the sociologists nor the equally exclusive biological determinism of the eugenicists and Social Darwinists. Human behavior and psychology, they said, was a complex interaction between biological and environmental factors at almost every stage of development.[23] From genomics and heritability, to the study of personality and IQ, nature and nurture had been shown to affect our behavior in interconnected ways--with some aspects of our behavior being affected more by biology and other aspects affected more so by our environment. Going further still, there are some behaviors, such as language, that are solely affected by the environment. And there are diseases, such as Huntington's disease, which are solely affected by genetics and heritability.[24]

In fact, sex is not solely controlled by *nature*, and gender is not solely controlled by *nurture*. When researchers discuss *sex differences* and *gender differences*, it is important to clarify what they mean. Sex differences are defined as sexually dimorphic (evolved sex differences). This could be physical characteristics such as average height. Gender differences, on the other hand, are defined as sexually monomorphic (same male-female) adaptations *with* culturally dimorphic socialization.[25] In other words, certain gender differences are more byproducts of sexually dimorphic adaptations, while other gender differences are more byproducts of socialization.

Therefore, the answer to the nature versus nurture debate is not "some of each" nor is it exclusively one or the other. Rather, the correct answer is *it depends on what traits are being discussed*, and this

---

[23] Moore, D. (2003). *The Dependent Gene: The Fallacy of "Nature vs. Nurture"*. Holt Paperbacks; M. Ridley, *Nature Via Nurture: Genes, Experience, and What Makes us Human*. Harper Collins.
[24] Pinker, *Why Nature and Nurture Won't Go Away*, 9.
[25] Mills, M. (2011). Sex Difference vs. Gender Difference? Oh, I'm So Confused! *Psychology Today*.

has implications on how we treat specific sex differences and specific gender differences on an aggregate level.

A Lesson on Bell Curves

Before we begin discussing aggregates of traits within a group and average group distributions, it is important to note what we mean when we say *aggregate* and *average*. In terms of aggregates, we are talking about the *sum total of the data sets* between two groups, not of individuals. An aggregate, similar to group averages, says nothing about an individual male or an individual female. While an aggregate assesses the sum total of the data, an average refers to the central value in the data. For example, human males have a higher aggregate (sum total) score of grip strength than human females. The average (central value) grip strength of a male is also higher than a female. Yet the aggregates and averages do not account for individuals. For example, there are some females who have higher grip strengths than some males. This concept can be understood using bell curve graphs. A normal bell curve such as the one presented in figure 3 only compares variations within a single group.

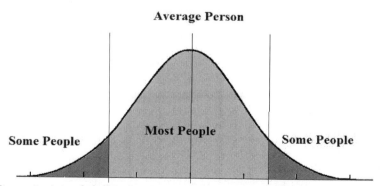

Figure 3. *A simple bell curve comparing within-group variation.*

However, a bimodal distribution (a double bell curve) compares variations within-groups *and* between-groups. Double bell curves are often seen when comparing two different groups of people, such as males and females. Depending on the variance of a specific trait, the double bell curve may overlap significantly or not overlap at all. Most between-group variations among males and females show significant overlap, meaning many males share characteristics with many females, and vice-versa. An oversimplified double bell curve shows this overlap in shared traits. While the average female and average male may not share the same degree of a trait, there is a significant overlap where individual males and females do not adhere to the average.

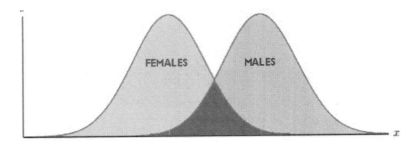

This simple caveat is important to remember as we delve into the discussion of aggregates and averages among groups of people. *The general rule is on an individual level to treat people as unique individuals, not as members of their group.* Yet in dealing with group statistics, we have to discuss differences using aggregates and averages.

Average Sex Differences

Most social constructionists claim that 1) *gender differences are social constructs with no basis in biology*, and 2) *disparities between the sexes are mostly attributable to sexism and societal-imposed gender roles.* I will address these claims by presenting recent evidence from the fields of psychology, biology, evolutionary behavioral science, neuroendocrinology, and neurology to show that gender differences are 1) real, and 2) largely attributable to biological factors, which affect average gender differences in interests and occupational preferences.

Let's first discuss physical differences. Average height difference between the sexes tends to be the most obvious, so we'll start with that. The average male is taller than the average female by 13 centimeters or almost half a foot, yet there is quite a large overlap between the sexes, as most people are in the 5 feet to 6 feet range. In this aspect, males and females seem to have about the same varying degrees of height, with males being taller on average, as seen in figure 4.

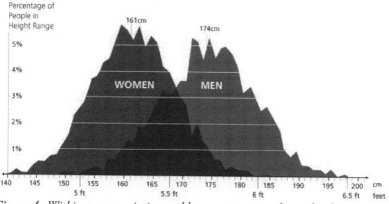

Figure 4. *Within-group variation and between-group overlap in height.*

Another physical difference is <u>average</u> combined grip strength. Once again, males have a higher average grip strength than the average female, yet there are quite a lot of females who have higher grip strengths than individual males, as seen in figure 5.

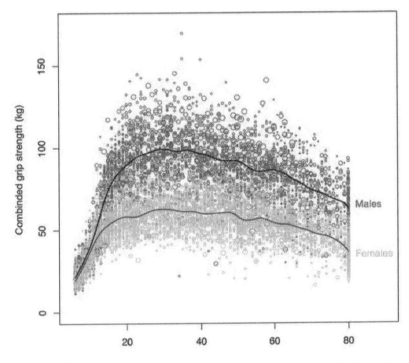

Figure 5. *Combined grip strength of males and females.* US National Health and Nutrition Examination Survey (NHANES) 2011-2012

Most average differences between males and females have quite a lot of overlap, which is good. However, there is one aspect of our bodies that feature almost no overlap, and that is our voice. The voice of the highest pitched male barely reaches the lowest pitched female, as seen in figure 6. Other major physical differences include upper body strength, muscle mass, face shape, width of the hips, running speed, throwing distance, and many others.

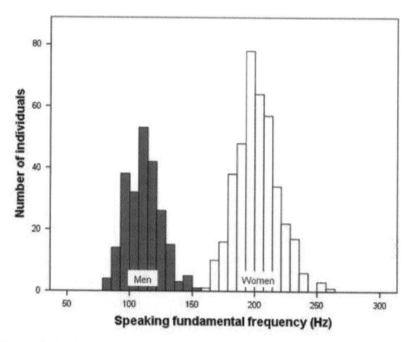

Figure 6. *Speaking fundamental frequency. Men in dark gray, women in white.* Puts et al. *Sexual Selection on Human Voices.* Evolutionary Psychology, 2014.

Almost no one has dared to argue that these clear physical differences are simply social constructions, and even though there are some radical sociologists who have, most accept these physical differences as biological factors arising from billions of years of natural selection. In fact, a key cause of physical differences between males and females stems from the amount of prenatal hormones the baby is subjected to in the womb, arguably the most important factor in shaping the male and female bodies as they grow into adulthood.[26]

Specifically, prenatal androgen plays a large role during gestation in determining sex differences and gender development.[27]

---

[26] Goldman, B. (2017). Two Minds: The Cognitive Differences between Men and Women. *Stanford Medicine Journal.*

[27] Hines, M. (2015). Early androgen exposure and human gender development. *Biology of Sex Differences, 6(3).*

During the second trimester, male brains are permanently altered in both function and structure by a sharp increase in androgen. This, in turn, produces more than just physical differences. It affects both psychological and physical traits, differentiating the two sexes psychologically and behaviorally.[28] As noted by Schmitt in *Psychology Today*, Prenatal androgen levels have been shown to predict six crucial things about development, thanks to multiple studies:

> "1) The degree of prenatal androgen exposure predicts differences of psychological traits in girls and boys.
>
> 2) Girls prenatally exposed to male-typical levels of androgens (compared to their unaffected sisters) express more male-typical psychology.[29]
>
> 3) Infants (as young as 5 months) exhibit psychological sex differences before extensive socialization.
>
> 4) Children exhibit many psychological sex differences before they have a conception of what sex/gender roles are or even what sex/gender is.
>
> 5) Experimental and observational studies of neurological and hormonal substrates of adult sexual identity, gender dysphoria, and transsexualism imply some degree of biological sexual differentiation in men's and women's psychology.

---

[28] Schmitt, D. (2016). Sex and Gender Are Dials (Not Switches). *Psychology Today*.
[29] Hines, M. (2016). Prenatal androgen exposure alters girls' responses to information indicating gender-appropriate behavior. *Philosophical Transactions, 371.*

6) Experimental and observational studies of nonhuman animals (including closely related primates) implicate evolved origins for many sex differences in personality, cognition and behavior."

In summary, androgen levels (amount of testosterone) during prenatal development have large effects on the physical, psychological, and behavioral traits of boys and girls before any socialization could occur.[30] Consider figure 7, which shows the average testosterone levels between males and females from conception through adulthood. The difference is large during the

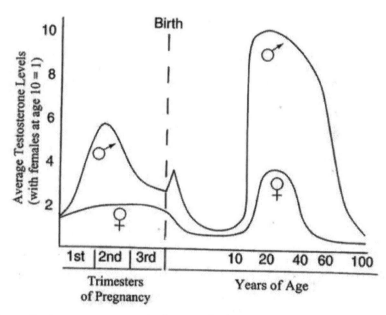

Figure 7. *Average testosterone levels across development between males and females.* Ellis, L. (2011). Identifying and Explaining Apparent Universal Sex Differences in Cognition and Behavior. *Personality and Individual Differences, 51*, 552-561.

---

[30] Schmitt (2016).

second trimester and incredibly large once puberty is reached. These levels indicate that males and females are incredibly different in terms of the type/amount of sex hormones, which causes many of the physical and psychological differences seen after puberty.

Differences in Cognition and the Brain

Decades of neuroscience research has shown males and females have, on average, have differing characteristics when it comes to cognition and how the brain functions, and some of this research has been done on animals. Recently, animal-research findings helped make a stronger case for the biological basis of sex differences in cognition. For example, a study of 34 rhesus monkeys showed that the male monkeys tended to prefer "toys with wheels over plush toys, whereas females found plush toys likable."[31] Did the monkeys' parents socialize them to like these different toys, or was this preference part of their biological predispositions? It would be pretty tough to argue for the former.

In terms of cognition in humans, males are adept at mental rotation and systematizing information, females are adept at mental location and verbal ability. *Once again, these are simply averages, not a description of all individuals.* Males, on average, have much lower reading comprehension and writing ability than females. Females, on average, are better at retrieving long-term memories, they are better at fine motor coordination and perceptual speed, and they excel at verbal ability over males.[32]

---

[31] Hassett, J. (2008). Sex Differences In Rhesus Monkey Toy Preferences Parallel Those Of Children. *Hormone Behavior, 54(3)*, 359-364.
[32] Goldman (2017).

From Silverman et al (2007), males, on average, are adept at visualizing and manipulating both 2D and 3D objects in their minds, at determining angles, and at tracking objects and projectiles.[33] Interestingly, navigation studies of both humans and rats showed that the females of both species relied more on landmarks to locate their position, while males tended to track their movements in the brain and estimate how far they've gone.

It is important to note that many of these differences, such as spatial-visualization, have been recorded in infants as young as two months.[34] Knowing all this, social constructionists need to ask themselves: Are these sex differences merely socially constructed or are they evidence of a larger evolutionary framework?

As it turns out, hormone manipulation at any critical stage can drastically affect these differences in cognition. For example, "Depriving newborn males of testosterone by castrating them or administering estrogen to newborn females results in a complete reversal of sex-typed behaviors in the adult animals. Treated females behave like males, and treated males behave like females."[35]

Though the brains of males and females are mostly similar, the neuroscience literature has a vast amount of research on the physical difference between male and female brains, showing that there is in fact a female brain and a male brain, with anatomical, structural, and physiological differences. A study in the Journal of Front Neuroendocrinology, published in 2010, documents the average sex differences of the brain in regards to neuroanatomy, neurochemistry,

---

[33] Silverman, I., Choi, J., and Peters, M. (2007). The Hunter-Gatherer Theory of Sex Differences in Spatial Abilities: Data from 40 Countries. *Archives of Sexual Behavior, 36*, 261-268.
[34] Goldman (2017).
[35] Kimura, D (1992). Sex Differences in the Brain. *Scientific American*, 119-125; 121.

and how male and female brains interact with specific medications that alter brain chemistry.[36] In January 2017, the Journal of Neuroscience Research released volume 95 of their journal, focused entirely on sex differences in the functioning of the male and female nervous systems.[37]

One of the most glaring differences between male and female brains is the amount of grey and white matter. Male brains tend to have nearly seven times more *grey matter* than female brains, while female brains tend to have ten times more *white matter* than male brains.[38] Grey matter functions as localized information processing centers that are involved in sensory perception and muscle control, and since male brains tend to have more, males exhibit more tunnel vision on intensive tasks than females. While grey matter is localized, white matter functions as the network. "White matter is the networking grid that connects the brain's grey matter and other processing centers with one another."[39] Because female brains tend to have more white matter, females exhibit a greater ability to switch quickly between tasks than males.

Regarding total brain size, the average male brain is bigger than the average female brain, while a female's hippocampus, which is involved in learning and memorization, is larger than a male's. The female brain has a higher density of neuron connections into the hippocampus, increasing sensory input. Areas of the brain that have larger size differentiation between males and females tend to have high amounts of receptors for sex hormones, something that makes

---

[36] Ngun, T. (2010). The Genetics of Sex Differences in Brain and Behavior. *Front Neuroendocrinology, 32(2)*, 227–246.
[37] Cahill, L. (2017). Sex/Gender Influences on Nervous System Function. *Journal of Neuroscience Research, 95*.
[38] Jantz, G. (2014). Brain Differences Between Genders. *Psychology Today*.
[39] *Ibid.*

sense if one understands the role testosterone and estrogen play in the male and female brains starting in the womb.[40]

In terms of mental health, males tend to have more cases of ADHD, retardation, and psychopathy, while females tend to have more cases of anxiety and depression.[41] A very recent study, analyzing 46,034 functional neuroimaging scans of 26,000 healthy patients showed that women, on average, have more active brains than men, demonstrating significant gender differences in regional cerebral blood flow. These findings can help with the specific treatment of psychiatric disorders and the understanding of neuroimaging itself.[42]

Other differences in a typical male and female physiology include the susceptibility to certain diseases and the ability to metabolize certain medications.[43]

Understanding all of these average differences helps medical professionals determine how a person's sex may affect the development of neurological disorders that "manifest and progress differently in men and women."[44] In other words, knowing these small but important sex differences in the brain improves the treatment of both men and women.

Another key to understanding the source of sex differences in the brain is understanding the effects of sex chromosomes, one of the 23 pairs of chromosomes in our cells. Biological males generally have one X and one Y chromosome, while biological females generally have two X chromosomes in each of their cells, including the brain. I say *generally* because there are rare instances where a biological male is

---

[40] Goldman (2017).
[41] Schmitt (2016).
[42] Amen, D. (2017). Gender-Based Cerebral Perfusion Differences in 46,034 Functional Neuroimaging Scans. *Journal of Alzheimer's Disease*, Preprint, 1-10.
[43] Ngun (2010).
[44] Ngun (2010).

given two X chromosomes or two X and two Y chromosomes. These are usually considered chromosomal disorders in the medical community, due to the abnormal and often life-altering symptoms associated with them. There are also individuals who are intersex, biological males or females who failed to fully develop their sexual organs, causing them to be ambiguous and consequently not fitting into the usual category of male or female. An intersex male might have an unusually small penis or missing/deformed testes, while an intersex female might have an unusually large clitoris or ambiguous ovaries. This is why when people discuss the fact they were "assigned" a gender, they are talking about the criteria doctors go through to determine your sex when you were born. Most of the population was born with normal and healthy sexual organs, clearly defining them within the male and female sexual category.

Nevertheless, the presence of these sex chromosomes (almost always XX or XY) affect the genetic traits and sex differences that will be expressed throughout development. And current neuroscience research is showing "direct genetic effects" on brain differences are significant, not excluding the role of prenatal androgens.[45] Furthermore, parts of the brain are now thought of as modules which are connected through neural pathways, and many of these modules, which are in charge of complicated behaviors, are masculinized or feminized through sex hormones.[46]

All these modules put together give you your "overall degree of maleness and femaleness."[47] There are individuals, of course, that depart from the average traits of masculine and feminine. A human

---

[45] *Ibid.*
[46] Yang, C. (2014). Representing Sex in the Brain, One Module at a Time. *Neuron*.
[47] Shah, N. (2016). Genetic Dissection of Neural Circuits Underlying Sexually Dimorphic Social Behaviors, *Royal Society Biological Sciences, 371(1688)*.

male, for example, may have a mix of traits that are more feminine than masculine, giving him a personality or temperament considered more feminine than the average male. The same is true for a human female whose traits add up to be more masculine than feminine, giving her a personality or temperament considered more masculine.

The level of masculine or feminine traits in a male and female is heavily tied to prenatal androgen levels and chromosomes, and all of these average differences in physical characteristics, cognition, neural structure, and the type/amount of sex hormones directly affect aspects of our personalities.[48]

---

[48] Beltz, A. (2011). Gendered Occupational Interests: Prenatal Androgen Effects on Psychological Orientation to Things versus People. *Hormone Behavior, 60(4)*, 313-317.

Differences in Personality

Both biological and social factors influence personality, and the degree to which specific traits are affected by either factor is sometimes difficult for researchers to gauge, but it continues to be an area of extensive research. Some of the traits that make up someone's personality are affected more by biology, while others are more affected by society or the environment. Studies that deal with these topics, especially cross-cultural studies, usually discuss this variance. Either way, it is clear that males and females, on average, have differing temperaments. These temperaments, or personalities, contribute to the interests we develop, which contribute to the choices we make and the occupations we gravitate towards as we become adults.

Most evolutionary psychologists and behavioral scientists agree that there are large gender differences in personality, as seen in cross-cultural studies such as Costa et al (2001) and Lippa et al (2008). Gender differences in personality are often discussed using averages so that scientists can assess the general differences between the sexes, while not excluding the nuance that some males score higher on feminine traits and some females score higher on masculine traits. Consider this quote from a paper in the *Frontiers in Psychology* journal.

> "The goal of investigating gender differences in personality, therefore, is to elucidate the differences among general patterns of behavior in men and women on average, with the understanding that both men and women can experience states across the full range of most traits. Gender differences in terms of mean differences do not imply that men and

women only experience states on opposing ends of the trait spectrum; on the contrary, significant differences can exist along with a high degree of overlap between the distributions of men and women."[49]

Many social constructionists become concerned when psychologists start to talk about sex and gender differences in personality, since these discussions often lead to debates about IQ and intelligence, and in history, these concepts were sometimes used as arguments for absurd ideas such as eugenics or the supposed "inferiority of women." Let's go ahead and put those fears to rest: Sex differences in general intelligence are negligible at best, and even specific areas of intelligence such as verbal and mathematical ability have negligible differences. No sex is superior or inferior to the other. We should also remember that *science only tells us what is and what is not*. It does not tell us how to act, or how to act morally. It is also important to clarify that no trait is ultimately better than another trait. *Every trait has its own positives and negatives.*

In the social constructionist view, due to the wide diversity of cultures and values, we would expect the personality differences between males and females to be widely different in cultures that differ from our own. For example, in Western society women on average have higher interests in professions that are artistic, social, and involve people, rather than men, who on average prefer professions that are orientated towards objects.[50] Yet in another

---

[49] Weisberg, Y. (2011). Gender Differences in Personality across the Ten Aspects of the Big Five. *Frontiers in Psychology, 2(178)*.
[50] Su, R., Armstrong, P. (2008). Men and Things, Women and People: A Meta-Analysis of Sex Differences in Interests. *Psychological Bulletin, 135(6)*, 859-884.

culture, who is to say that these average interests won't be flipped? Let's take a look at the research.

Over the past three decades, multiple meta-analysis studies have been conducted regarding average gender differences in personality traits, using self-report personality questionnaires to gather the data. A few are worth noting. (Meta-analysis is a large analysis of multiple research studies in one single study, allowing researchers to obtain a large amount of data and analyze it all in a single data pool.) Feingold's meta-analysis study in 1994 of the gender differences in personality, Costa's 2001 study regarding gender differences in personality traits across cultures, Lippa's 2008 cross-cultural study on gender differences in interests, and Ellis's 2011 cross-cultural study on universal sex differences all provide evidence for the idea that average sex-gender differences in personality traits are consistent across cultures.[51]

The most popular measurement tool of personality dimensions in psychology is known as the *Big Five*, which categorizes personality into five traits: Extraversion, Agreeableness, Neuroticism, Conscientiousness, and Openness to Experience. Each of the five major traits in the *Big Five* has two subcategories which are separate but also somewhat correlated, conceptualizing the framework as a kind of hierarchy with five major traits and ten sub-traits. For example, Extraversion's sub-traits are *enthusiasm* and *assertiveness*, while the sub-traits for Agreeableness are *compassion* and *politeness*.[52]

Researchers collecting data between males and females often use effect size (d value) to quantify the difference in the averages

---

[51] Costa, P. T., Jr., Terracciano, A., & McCrae, R. R. (2001). Gender Differences in Personality Traits across Cultures: Robust and Surprising Findings. *Journal of Personality and Social Psychology, 81(2)*, 322-331.

[52] Weisberg (2011).

(remember the double bell curve?). A positive value of +0.50 means males score moderately higher than females in a specific trait, while a negative value of -0.50 means females score moderately higher than males in a specific trait.[53]

On a broad scale, males and females are fairly similar psychologically speaking, with a large amount of overlap. Yet some traits between males and females show more distinctions than others. In Feingold's meta-analysis, gender differences in trust were found to be small (d = -0.20), indicating that 58% of women are higher in trust than the average man. Gender differences in tender-mindedness (associated with trait Agreeableness) were found to be large (d = -1.00), indicating that 84% of women are higher in tender-mindedness than the average man, signifying this trait may be affected more by biological factors.[54]

Weisberg's 2011 study on personality differences across the *Big Five* provided similar data. For example, while the difference between the sexes in the traits of Openness/Intellect and Conscientiousness was almost zero (d = 0.02; d = -0.06), the *compassion* sub-trait of Agreeableness was found to be moderately higher in the average woman than the average man at d = -0.45. In Weisberg's study, no effect size among the traits on the Big Five was found to exceed 0.48, indicating that there tends to be no major personality trait in which the sexes differ at an extremely high level. This means most of the double bell curves for personality differences in males and females overlap significantly, and most of these trait averages stay relatively consistent across culture and time. Even though these differences

---

[53] Schmitt, D. (2015). How Big are Psychological Sex Differences? *Psychology Today*.
[54] Feingold, A. (1994). Gender Differences in Personality: A Meta-analysis. *Psychological Bulletin, 116(3),* 429-456.

tend to be moderate, they still have an effect on individual interests and behavior.

There is one aspect of temperament that has one of the largest effect sizes in personality, and that trait is physical aggressiveness. In a 2004 study, levels of physical aggression between the sexes were found to have an effect size of +0.80, indicating that 79% of men are higher than the average woman in this aspect, which is not surprising, knowing the much higher level of androgens in males.[55] This high effect size, along with the fact that it stays consistent across cultures, indicates that physical aggression is sexually dimorphic, not culturally dimorphic.

Lippa's study focused on gender differences in personality and occupational preferences across cultures, which tested both the biological and social-environmental theories of gender. Specifically, the personality traits of Extraversion, Agreeableness, Neuroticism (tendency to experience negative emotion), and male-versus-female-typical occupational preferences (MF-Occ) were examined, using 200,000 participants across 53 nations. The male and female participants differed significantly in the four traits across cultures. Their results further substantiated the view of most psychologists and evolutionary biologists: *Males and females differ in terms of average personality traits, even across different cultures.* For example, Extraversion, Agreeableness, and Neuroticism, tend to be higher in females than in males, with females varying more in extraversion and males varying more in agreeableness.[56] Sex differences in emotionality such as trait Neuroticism depend on the type of emotion and other

---

[55] Archer, J. (2004). Sex Differences in Aggression in Real-World Settings: A Meta-Analytic Review. *Review of General Psychology, 8,* 291-322.

[56] Lippa, R. (2008). Sex Differences in Personality Traits and Gender-Related Occupational Preferences across 53 Nations. *Sexual Behavior, 39,* 619-636.

factors, as males tend to be more emotional in certain aspects, while females tend to be more emotional in other aspects.[57] All of these averages suggest that biological factors play a large role in shaping interests and personality more so than cultural or societal factors, with sociocultural factors playing a smaller role.[58]

Ellis's cross-cultural study in 2011 further documented the specific psychological differences between males and females. Inconsistent with the social constructionist hypothesis, the researchers found 65 universal sex differences in terms of psychology and behavior across the world.[59] In other words, it didn't matter what culture you went to, there were many traits between males and females that remained fairly different, indicating that many personality traits are sexually dimorphic and culturally monomorphic (evolved sex differences). Perhaps specific personality traits that show strong signs of sexual dimorphism should be considered innate *sex differences*, not gender differences.

Costa's 2001 study had a surprising finding that provides important insights into personality research: The view that average gender differences would be consistent across cultures was found to be slightly true and slightly false. While most gender differences across cultures do exist, the *magnitude* of these differences vary across cultures. In other words, while all cultures have degrees of gender differences, some cultures have a higher degree than others. The nations with higher degrees of gender differences were predominantly Western, non-patriarchal liberal nations: "Contrary to predictions

---

[57] Schmitt, D. (2017). On That Google Memo about Sex Differences. *Psychology Today*.
[58] Beltz (2011).
[59] Ellis, L. (2011). Identifying and Explaining Apparent Universal Sex Differences in Cognition and Behavior. *Personality and Individual Differences, 51*, 552-561.

from evolutionary theory, the magnitude of gender differences varied across cultures. Contrary to predictions from the social role model [social constructionism], gender differences were most pronounced in European and American cultures in which traditional sex roles are minimized."[60] This finding, that sex differences grow larger with increased gender equality, is called the *Gender Equality Paradox*:

> The degree of personality differences between males and females is <u>positively correlated</u> with the degree of gender equality in a nation, meaning **the more gender equal and egalitarian a nation is, the larger the gender differences in personality are.**

> In other words, **gender differences in personality actually *maximize* in Western, egalitarian nations, especially those which have high economic opportunity, as opposed to nations which are non-egalitarian and have low economic opportunity.** "Recent studies have shown that many gender differences in personality tend to be larger in more developed, Western cultures with less traditional sex roles."[61] Some of these studies include the Costa 2001 study and the Schmitt et al 2008 study across 55 cultures, providing evidence for the large role biology may play in affecting temperament and personality.[62]

---

[60] Costa et al (2001).
[61] Weisberg (2011).
[62] Schmitt, D. (2008). Why can't a man be more like a woman? Sex differences in Big Five personality traits across 55 cultures. *Journal of Personality and Social Psychology, 94(1),* 168-182; Costa et al (2001).

To many psychologists and sociologists who had studied the effects of society and culture on gender, this was unintuitive. In fact, they were proposing the *opposite* would happen.

Think about this. If the social constructionists are correct, as nations become more egalitarian, gender differences should minimize, since gender is simply a sociocultural construct. Yet this is not what we find. In nations like Denmark, Sweden, and Norway (the nations where gender equality is highest according to the UN), personality differences between males and females are much larger than patriarchal nations like Pakistan, Saudi Arabia, or Egypt. Furthermore, even sex differences in mental rotation ability and line angle judgments were larger in gender egalitarian countries, with these abilities increasing for both sexes as countries became more equal.[63]

If these studies on personality are correct, it would mean that gender differences in personality, and consequently occupational preferences, are heavily affected by biological factors, since these egalitarian nations have successfully minimized most sociocultural factors that might limit or affect freedom of choice or economic opportunity.[64]

There are three major views for the sources of these differences. One is the view of many psychologists, biologists, and evolutionary biologists: "Biological and evolutionary approaches posit that gender differences are due to men and women's dimorphically evolved concerns with respect to reproductive issues, parental investment in

---

[63] Lippa, R. A., Collaer, M. L., & Peters, M. (2010). Sex differences in mental rotation and line angle judgments are positively associated with gender equality and economic development across 53 nations. *Archives of Sexual Behavior, 39*, 990-997.
[64] Schmitt (2017).

offspring."⁶⁵ This view emphasizes the importance of the biological over the sociocultural in shaping male and female actions.

On the other hand, social constructionists tend to argue that males and females are socialized by society to behave differently than one another, seeing biological factors as minimal or non-existent.⁶⁶

The third view is that biological and societal/environmental factors play a role to differing degrees across different traits, and this tends to be the view held by most scholars today.

The tendency of some traits to not have any cultural consistency while others do is interesting. This suggests that some traits may be heavily influenced by culture and society (culturally dimorphic) while others may be more heavily influenced by evolutionary and biological factors (sexually dimorphic). It also requires that we recognize the complicated and detailed nature of the origins of personality. Weisberg, in his review of the literature, concludes, "Exactly how culture impacts personality is a complex question, worthy of future study."

In summary, personality differences are affected to a certain degree by biological factors, as evidenced by the studies of larger personality differences in egalitarian nations compared to non-egalitarian ones, as well as the cross-cultural studies mentioned previously. It is also true that some traits exhibit evidence of societal origins. On the broad scale, personality differences between the sexes are less pronounced compared to their physical differences, and regardless of their origin, personality differences between the sexes do

---

[65] Weisberg (2011).
[66] Eagly A. H., Wood W. (2005). Universal Sex Differences across Patriarchal Cultures ≠ Evolved Psychological Dispositions. *Behavioral Brain Science, 28*, 281–283.

exist to a certain degree, and these differences are directly linked to the variations we see in interests and occupational preferences.

Differences in Interests and Occupational Preferences

Disparities across occupations (such as men being the majority in engineering fields, and women being the majority in social sciences) can be largely attributed to sex differences in average interests (i.e. men and women preferring different types of work).[67] These differences in average interests and occupational preferences are partially linked to biology, partially linked to the social environment, and partially linked to personality, as many studies have shown.

In 2008, psychologists at the University of Illinois at Urbana–Champaign and Iowa State University conducted the first ever meta-analysis on sex differences in vocational interests, with over *500,000* participants, an unheard of number in psychological studies. Their goal was to examine sex differences in interests on a broad scale. In the most simplified terms, what they found was that <u>men preferred working with things and women preferred working with people</u>.[68] *Remember the rhesus monkeys?* The researchers found the effect size to be d = 0.93 on the Things-People dimension, one of the largest effect sizes in psychology.[69] And this difference could not be more

---

[67] Su et al (2008); Lippa (2010); Jussim (2017); Wang at el (2013); Schmitt (2017); Beltz (2011).

[68] Su et al (2008); Lippa, R. (1998). Gender-Related Individual Differences and the Structure of Vocational Interests: The Importance of the People–Things Dimension. *Journal of Psychology and Social Psychology, 74(4)*, 996-1009.

[69] Jussim, L. (2017). Why Brilliant Girls Tend to Favor Non-STEM Careers. *Psychology Today*.

pronounced, indicating that only 15% of women have the same level of interest as the average man in fields like engineering.[70] It also means the bimodal distribution curves are farther apart on this dimension, which also means that there is less overlap between men and women in this aspect. Researchers concluded in another similar study from 2010 that biological sex "accounted for 33% of the variance in occupational interests," with this variance becoming larger as the United States became more egalitarian.[71] These studies show that many occupational disparities are heavily linked to average sex differences in interests.

A secondary goal of the researchers was to provide sufficient information on "the size and pattern of sex differences in interests" so that their data could help inform both researchers and policy-makers. Their large participant volume along with the large amount of data contributed a significant amount of information into the discussion. The researchers concluded with some interesting points:

> "These results indicate the important role of interests in gendered occupational choices and gender disparity in the STEM fields and have implications for how future interest measures are developed and used in applied settings…These sex differences are remarkably consistent across age and over time, providing an exception to the generalization that only small sex differences exist…Educators and counselors also need to be careful in choosing assessment tools and in interpreting the results of such measures so as not to restrict

---

[70] Su et al (2008).
[71] Lippa, R. A. (2010). Sex differences in personality traits and gender-related occupational preferences across 53 nations: Testing evolutionary and social-environmental theories. *Archives of Sexual Behavior, 39,* 619–636.; Schmitt (2017).

the occupational choice of individuals—for both men and women."[72]

Yet even though there is still a large gender disparity in STEM, there are ways of reducing this through socialization especially at a young age, say the researchers.

Another study from 2011 focused on differences in interests and how they related to math and verbal ability using 1,490 high school students across the nation.[73] The researchers concluded with four interesting findings. *Psychology Today* reported the results.

"1) 70 percent more girls than boys had strong math *and* verbal skills.

2) Boys were more than twice as likely as girls to have strong math skills but not strong verbal skills.

3) People (regardless of whether they were male or female) who had only strong math skills as students were more likely to be working in STEM fields at age 33 than were other students.

4) People (regardless of whether they were male or female) with strong math and verbal skills as students were less likely to be working in STEM fields at age 33 than were those with only strong math skills."[74]

---

[72] Su et al (2008).
[73] Wang, Eccles, and Kenny (2013). Not Lack of Ability but More Choice: Individual and Gender Differences in Choice of Careers in Science, Technology, Engineering, and Mathematics. *Psychological Science, XX(X)*, 1-6.
[74] Jussim (2017).

Notice that those with only strong math skills were more likely to work in STEM, while those with both strong math skills and verbal skills were less likely to be working in STEM. The authors hypothesize that this difference has to do with the degree of self-concept regarding math ability. If a person views their math skills as strong but their verbal skills as moderate, they are more likely to have a higher math-ability self-concept, as opposed to a person who views their math and verbal ability as equally strong. This higher math-ability self-concept (focus on math skills alone) tends to lead people towards STEM fields. Males, on average, tend to have higher math-ability self-concepts, even though their female counterparts may be just as good at math. Perhaps the key to more women in STEM is increasing their math-ability self-concepts--and social factors can certainly have a role to play in this strategy. Since we know that sex differences in *math ability, achievement, and performance are very small or negligible* (note the equal percentages of male and female bachelor's degree holders in Mathematics) [75], the large STEM disparity is not due to ability. Part of it has to do with the level of self-concept and average sex differences in interest.[76]

The degree to which societal or biological factors have affected the interests of these women cannot be entirely certain, but what is certain is the *existence of the large sex difference in occupational preferences.*

---

[75] Lindberg, S., Hyde, J. (2010). New trends in gender and mathematics performance: A meta-analysis. *Psychological Bulletin, 136(6).*
[76] Su et al (2008); Lippa (1998).

Average Sex Differences: Summarized

From physical characteristics and hormone levels, to cognition, brain size, personality, and interests, the average differences between males and females across the world do exist and are distinct--with some varying degrees of difference.

Physical characteristics, specifically characteristics about the male and female bodies, is arguably the one aspect of men and women which is the most pronounced across cultures and is almost entirely due to biological and environmental factors. These factors do not relate to socialization. Aspects of physical characteristics such as androgen levels, average height, grip strength, muscle mass, vocal pitch, upper body strength, running speed, and throwing distance have high aggregate differences. And these differences are very much related to the presence and amount of androgens in the body, both prenatally and during puberty.[77] Most physical characteristics, therefore, are *sexually dimorphic adaptations.*

Hormone level differences are usually very consistent, with some outliers. Girls who are prenatally exposed to male levels of androgen tend to exhibit more masculine psychological traits, while boys who are prenatally exposed to lower than normal male levels of androgen tend to exhibit more feminine psychological traits. Furthermore, very young infants exhibit psychological sex differences before socialization even occurs, indicating that sex hormones have a large effect on shaping the psychology of males and females.[78] Hormone levels, therefore, are *sexually dimorphic adaptations.*

Cognition differences between the sexes tend to be less pronounced than differences in physical characteristics, yet they are

---

[77] Schmitt (2016).
[78] *Ibid.*

still statistically significant. While males tend to be more adept at mental rotation and systematizing information, females tend to be more adept at mental location and verbal ability. Males, on average, have much lower reading comprehension and writing ability than females. Females, on average, are better at retrieving long-term memories, they are better at fine motor coordination and perceptual speed, and they excel at verbal ability over males.[79] Males are *slightly* more adept at spatial-visualization than the average female.[80] These averages have been observed in infants as young as two months, indicating that cognition differences are predominantly *sexually dimorphic adaptations.*

Brain differences are very prominent, including anatomical, structural, and physiological differences. The average male brain is bigger than the average female brain, while a female's hippocampus, which is involved in learning and memorization, is larger than a male's. It also works differently, and the same is true for the amygdala.[81] Male brains tend to have nearly seven times more *grey matter* than female brains, while female brains tend to have ten times more *white matter* than male brains.[82] Differences in psychiatric disorders are also present, with males suffering from higher rates of ADHD, retardation, and psychopathy than females, while females suffering from higher rates of anxiety and depression than males.[83] Moreover, regions of cerebral blood flow showed large sex differences across 46,000 neuroimaging scans, with female brains tending to be more active.[84]

---

[79] Goldman (2017).
[80] Silverman et al (2007).
[81] Goldman (2017).
[82] Jantz (2014).
[83] Schmitt (2016).
[84] Amen et al (2017).

These differences in brain anatomy, mental health, and structure are mostly attributable to genetic and evolutionary factors, not socialization. Mental health may be the one exception here. Generally speaking, the difference in the brain's structure and anatomy between the sexes is mostly caused by *sexually dimorphic adaptations.*

Personality is complicated. Differences between males and females in this category are prominent, but they are not as large as differences in physical characteristics, differences in hormone levels, or differences in brain size. *Out of any category, personality seems to have the highest overlap between males and females,* indicating personality is incredibly variable, while still being affected by both biological and societal factors to differing degrees.

First, on the cultural level: Feingold's meta-analysis in 1994 of the gender differences in personality, Costa's 2001 study regarding gender differences in personality traits across cultures, and Lippa's 2008 cross-cultural study on gender differences in interests all provide evidence for the universality of average sex differences in personality traits.

Specifically, females on average score higher in Extraversion, Agreeableness, and Neuroticism than their male counterparts. These average scores remain higher in females than in males across cultures.[85]

Second, on the effect size: Effect sizes in personality psychology seem to stay relatively moderate, with most traits staying within the 0.2-0.5 range. This indicates that there is a significant overlap of personality traits. For example, effect size for trait Conscientiousness and trait Openness/Intellect was found to be almost zero, meaning most males and females share equal levels of these traits.[86] Traits such

---

[85] Lippa et al (2008).
[86] Weisberg (2011).

as Agreeableness and Extraversion, while higher for females, are only slightly higher, meaning the proportional difference in the population is about 60% female to 40% male in these traits. Many personality traits have this large overlap.

Third, on the correlation between gender equality and gender differences: While some gender differences have decreased as countries have become more equal, multiple studies found that many sex differences in personality actually become larger in Western egalitarian nations.[87] **The degree of personality differences between males and females is positively correlated with the degree of gender equality in a nation.**[88] Data on personality differences was collected from 55 countries to arrive at this conclusion, indicating biology may play a large role in average personality differences between the sexes.

As I have discussed, the social constructionist viewpoint predicts the opposite of this. Theoretically, gender differences should minimize to almost zero as gender equality increases. But that's not what happens.

Conclusion for personality: *Some traits are sexually dimorphic, while others are culturally dimorphic.* Biological factors that affect personality include prenatal androgen levels, brain structure, and cognition. Societal factors that affect personality include the family environment, socialization, peer-to-peer interaction, and specific life experiences.

<u>**Interests and occupational preferences**</u> seem to be tied to five things: 1) physical characteristics 2) the way an individual's brain works (cognition), 3) personality, 4) socialization, and 5) life experiences. 1 and 2 are mostly biological factors, 3 is more balanced,

---

[87] Costa et al (2001).
[88] Schmitt et al (2008).

and 4 & 5 are, of course, societal/environmental. Physical characteristics tend to affect jobs that need specific physical characteristics, such as mining, drilling, operating heavy machinery, construction work, football, serving in the frontlines of a military. Much more males dominant these areas due to high aggregate differences in physical characteristics. Both cognition and personality have already been shown to be heavily affected by biology (with personality being more nuanced), and these two aspects heavily affect interests later in life. Socialization and life experiences can also affect the decisions an individual makes, and these should be accounted for when discussing occupational preferences as well.

A meta-analysis performed in 2008 with over 500,000 participants showed that men prefer working with things and women prefer working with people. The effect size was d = 0.93 in the Things-People dimension, and this average can explain a large percentage of occupational disparities in gender egalitarian nations.[89] This indicates that interests are largely *sexually dimorphic*, relying on genetic factors and a portion of environmental factors.

Lastly, even though males and females have equal performance levels in math, levels of both self-concept and overall interest may play a role in whether males and females choose to pursue STEM careers.[90]

---

[89] Su et al (2008); Lippa (1998); Beltz (2011).
[90] Wang et al (2013).

Conclusions

Inconsistent with the hypothesis that most or all gender differences are the result of social constructions, a review of the evidence from the fields of psychology, biology, evolutionary behavioral science, neuroendocrinology, and neurology (Goldman 2017, Hines 2016, Hines 2015, Wang et al 2013, Beltz 2011, Ellis et al 2011, Weisberg 2011, Su et al 2008 meta-analysis with 500,000 people, Schmitt et al 2008, Lippa et al 2008, Silverman et al 2007, Archer et al 2004, Costa et al 2001, Lippa 1998, Feingold et al 1994, and many more) **suggests that gender differences are 1) real and 2) affected by biological factors.** Many of these sexually dimorphic traits are even seen in infants and across cultures, and these traits differentiate the sexes *on average* in terms of cognition and personality, **leading to average differences in interests and occupational preferences.**[91]

Gender, as it turns out, is not simply a social construct. While it is true that gender identity (sense of maleness or femaleness) is heavily influenced by societal factors[92], most of the population identifies with their biological sex, indicating that gender is tightly correlated with biology.[93] In fact, for almost the entire species, biological sex, gender/gender identity, and sexual orientation are all heavily dependent on each other.[94] Altering one can substantially affect the other. Furthermore, while there are some gender traits that differ across cultures, most of the gender differences in traits have been

---

[91] Goldman (2017).
[92] Moss-Racusin (2010). When Men Break the Gender Rules: Status Incongruity and Backlash Against Modest Men. *Psychology of Men & Masculinity, 11*, 140–151.
[93] Ngun (2010).
[94] Ludden, D. (2016). When Sex and Gender Don't Match. *Psychology Today*.

proven universal across cultures and time, indicating that many of these gender differences may actually be *sex differences* to a certain degree.[95]

One important conclusion from this body of evidence is that each human being is incredibly variable and unique. Aggregates and averages of male and female group distributions say nothing about an individual--whether an individual may prefer a certain type of job over another or whether an individual will have higher scores of trait Agreeableness or trait Extraversion. There is significant overlap between men and women in many traits. In fact, no male is fully masculine, and no female is fully feminine. We all have differing degrees of masculine and feminine traits. Some males may have higher degrees of feminine traits than masculine traits, and vice-versa. That's why we have personality tests in the first place!

Secondly, it should now be clear why gender disparities in egalitarian countries such as the United States still exist, and why they continue to be larger in these countries than in others, such as the disparities in STEM fields.

The evidence indicates these inequalities are mostly the result of innate sexually dimorphic traits, leading people to choose different career paths. Specific reasons for these preferences can be explained through prenatal androgen levels that cause differences in cognition and personality, and direct genetic effects.[96] Socialization factors also play a role, such as levels of self-concept, the influence of the family environment, peer-to-peer interaction, and teachers.[97]

---

[95] Lippa et al (2008); Costa et al (2001); Feingold et al (1994).
[96] Ngun (2010).
[97] Wang et al (2013).

For example, does the inequality in the health sciences mean the field is discriminating against men? No, there is no evidence to suggest that the inequality in this field is the result of discrimination. Rather, more percentages of men prefer studying business, physical/earth sciences, math/computer sciences, and engineering, rather than the health sciences. And the reasons for these disparities are clear.

Simply put, the decisions men and women make in gender egalitarian countries such as the United States, Norway, Denmark, Sweden, and Germany have major biological and environmental causes that do not relate to sexism or discrimination. (Individual instances of discrimination still exist, but they are a rarity, not the norm in these countries. Instances of discrimination should be shown through clear evidence and dealt with accordingly.)

Consider another important conclusion: *inequalities do not necessarily equal injustice.* **The continuation of these inequalities in gender egalitarian countries** (nations that have minimized many social factors that may affect choice) **is mostly the result of average differences in interests.**[98] This is particularly true in Western liberal nations, where individuals have high levels of economic opportunity. As long as nations have high levels of economic and social freedom, inequalities will continue to exist in differing degrees. (What matters is the level of happiness and standard of living.) Averages of people's choices, preferences, and interests must be taken into account when evaluating the source of these inequalities. (Think about the earnings gap!) Disparities seen among occupations and other distributions can be explained through differing interests, differing personalities,

---

[98] Su et al (2008); Schmitt et al (2008); Costa et al (2001).

differing neural structure, differing cognitive abilities, and sometimes (as is the case in hard labor jobs) differing physical characteristics.[99]

This does not mean that "women are not suited for tech jobs" or "men are not suited for social work."[100] Rather, **both men and women, on average, have <u>natural inclinations to prefer specific types of jobs.</u>** It doesn't mean they are "not suited." There are many women who excel in engineering compared to other men. There are many men who excel as therapists compared to other women. If a woman enjoys computer science, she should pursue those interests, and if a man enjoys elementary education, he should pursue those interests as well. *Both men and women simply want to be in an occupation that will make them happy.*

Knowing these average differences between the sexes can help us develop more effective strategies for increasing women's interest in STEM and maybe even increasing men's interest in the social sciences and humanities--strategies that do not involve quotas and discriminatory hiring practices. Based on many studies, one of the reasons women may not pursue STEM careers is due to their levels of self-concept in math-ability. Improving these levels in girls, through social influences like teachers, parents, peers, and mentors, is one possible way of increasing women's interest in these fields, and it could be especially effective for girls who already excel at math.

All of this evidence, from the effects of prenatal androgen in psychology and the differences in cognition observed in infants, to the differences in brain structure, personality, and interests, is

---

[99] Weisberg (2011).
[100] Wootson, C. (2017). A Google engineer wrote that women may be unsuited for tech jobs. Women wrote back. *Washington Post*.

inconsistent with social constructionism.[101] And the evidence is vast. **Average biological, psychological, and behavioral differences between human males and human females is significant.** While men and women have much more in common than different, understanding that average differences between the sexes do exist, and that these differences are often good and healthy, will surely improve our society's medical, mental, and sexual health.[102] As the years pass, more and more evidence will continue to clarify the degree biological factors and societal factors affect specific traits, including people's interests and preferences.

And as this evidence becomes clearer, the principles of Postmodernist social constructionism will continue to become weaker. The beliefs in *tabula rasa,* social determinism, perfect gender equity, and a genderless utopia will be proven outdated and misguided, based on a dying 20th century philosophy.[103] Yet the Enlightenment values of objective truth, individualism, liberty, reason, and science will continue to triumph as liberal nations move through the 21st century.

---

[101] Goldman (2017); Hines (2016); Hines (2015); Wang et al (2013); Ellis (2011); Su et al (2008); Silverman et al (2007).
[102] Ngun (2010).
[103] Hicks (2004).

References

Amen, D. (2017). Gender-based cerebral perfusion differences in 46,034 functional neuroimaging scans. *Journal of Alzheimer's disease*, Preprint, 1-10.

Archer, J. (2004). Sex differences in aggression in real-world settings: a meta-analytic review. *Review of General Psychology, 8*, 291-322.

Beltz, A. (2011). Gendered occupational interests: prenatal androgen effects on psychological orientation to things versus people. *Hormone Behavior, 60(4)*, 313-317.

Cahill, L. (2017). Sex/gender influences on nervous system function. *Journal of Neuroscience Research, 95.*

Carson, A. (2013). Prisoners in 2013. *Department of Justice, Bureau of Justice Statistics.*

Coleman, N. (2017). Trans man gives birth to baby boy. *CNN.*

Costa, P. T., Jr., Terracciano, A., & McCrae, R. R. (2001). Gender differences in personality traits across cultures: robust and surprising findings. *Journal of Personality and Social Psychology, 81(2)*, 322-331.

Derrida, J. (1994). *Specters of Marx*. Routledge.

Eagly A. H., Wood W. (2005). Universal sex differences across patriarchal cultures ≠ evolved psychological dispositions. *Behavioral Brain Science, 28*, 281–283.

Ellis, L. (2011). Identifying and explaining apparent universal sex differences in cognition and behavior. *Personality and Individual Differences, 51*, 552-561.

Feingold, A. (1994). Gender differences in personality: a meta-analysis. *Psychological Bulletin, 116(3),* 429-456.

Fisher, A. (2013). Boys vs. Girls: What's Behind the College Grad Gender Gap? *Forbes.*

Goldman, B. (2017). Two minds: the cognitive differences between men and women. *Stanford Medicine Journal.*

Hassett, J. (2008). Sex differences in rhesus monkey toy preferences parallel those of children. *Hormone Behavior, 54(3),* 359-364.

Hicks, S. (2004). *Explaining Postmodernism: Skepticism and Socialism from Rousseau to Foucault.* Scholargy Publishing, Inc.

Hines, M. (2015). Early androgen exposure and human gender development. *Biology of Sex Differences, 6(3).*

Hines, M. (2016). Prenatal androgen exposure alters girls' responses to information indicating gender-appropriate behavior. *Philosophical Transactions, 371.*

Holmes, M. (2007). *What is Gender? A Sociological Approach.* Sage Publications.

Horwitz, S. (2017). Truth and Myth on the Gender Pay Gap. *Foundation for Economic Education.*

Jantz, G. (2014). Brain Differences Between Genders. *Psychology Today.*

Jussim, L. (2017). Why Brilliant Girls Tend to Favor Non-STEM Careers. *Psychology Today.*

Kimura, D. (1992). Sex Differences in the Brain. *Scientific American,* 119-125.

Lindberg, S., Hyde, J. (2010). New trends in gender and mathematics performance: A meta-analysis. *Psychological Bulletin, 136(6).*

Lippa, R. (1998). Gender-related individual differences and the structure of vocational interests: the importance of the people–things dimension. *Journal of Psychology and Social Psychology, 74(4),* 996-1009.

Lippa, R. (2008). Sex differences in personality traits and gender-related occupational preferences across 53 nations. *Sexual Behavior, 39,* 619-636.

Lippa, R. A., Collaer, M. L., & Peters, M. (2010). Sex differences in mental rotation and line angle judgments are positively associated with gender equality and economic development across 53 nations. *Archives of Sexual Behavior, 39,* 990-997.

Lippa, R. A. (2010). Sex differences in personality traits and gender-related occupational preferences across 53 nations: Testing evolutionary and social-environmental theories. *Archives of Sexual Behavior, 39,* 619–636.

Ludden, D. (2016). When Sex and Gender Don't Match. *Psychology Today.*

Mills, M. (2011). Sex Difference vs. Gender Difference? Oh, I'm So Confused! *Psychology Today.*

Moore, D. (2003). *The Dependent Gene: The Fallacy of "Nature vs. Nurture'.* Holt Paperbacks; M. Ridley, *Nature Via Nurture: Genes, Experience, and What Makes us Human.* Harper Collins.

Moss-Racusin (2010). When men break the gender rules: status incongruity and backlash against modest men. *Psychology of Men & Masculinity, 11,* 140–151.

Ngun, T. (2010). The genetics of sex differences in brain and behavior. *Front Neuroendocrinology, 32(2),* 227–246.

Nichols, J. (2017). Women Aren't The Only Ones Who Get Periods. *HuffPost.*

Pinker, S. (2004). Why Nature and Nurture Won't Go Away. *Daedalus 133(4)*, 9.

Schmitt, D. (2008). Why can't a man be more like a woman? Sex differences in Big Five personality traits across 55 cultures. *Journal of Personality and Social Psychology, 94(1)*, 168-182.

Schmitt, D. (2015). How Big are Psychological Sex Differences? *Psychology Today*.

Schmitt, D. (2016). Sex and Gender Are Dials (Not Switches). *Psychology Today*.

Schmitt, D. (2017). On That Google Memo about Sex Differences. *Psychology Today*.

Shah, N. (2016). Genetic dissection of neural circuits underlying sexually dimorphic social behaviors. *Royal Society Biological Sciences, 371(1688)*.

Silverman, I., Choi, J., and Peters, M. (2007). The hunter-gatherer theory of sex differences in spatial abilities: data from 40 countries. *Archives of Sexual Behavior, 36*, 261-268.

Stoet, G., and Geary, D. C. (2015). Sex differences in academic achievement are not related to political, economic, or social equality. *Intelligence, 48*, 137-151.

Su, R., Armstrong, P. (2008). Men and things, women and people: a meta-analysis of sex differences in interests. *Psychological Bulletin, 135(6),* 859-884.

Wang, Eccles, and Kenny (2013). Not lack of ability but more choice: individual and gender differences in choice of careers in science, technology, engineering, and mathematics. *Psychological Science, XX(X),* 1-6.

Weisberg, Y. (2011). Gender differences in personality across the ten aspects of the Big Five. *Frontiers in Psychology, 2(178).*

Wootson, C. (2017). A Google engineer wrote that women may be unsuited for tech jobs. Women wrote back. *Washington Post.*

Yang, C. (2014). Representing sex in the brain, one module at a time. *Neuron.*

Made in the USA
Monee, IL
04 September 2023

42136866R00042